U.S. ENVIRONMENTAL PROTECTION AGENCY
OFFICE OF INSPECTOR GENERAL

Evaluation Report

EPA's Office of Research and Development Performance Measures Need Improvement

Report No. 10-P-0176

August 4, 2010

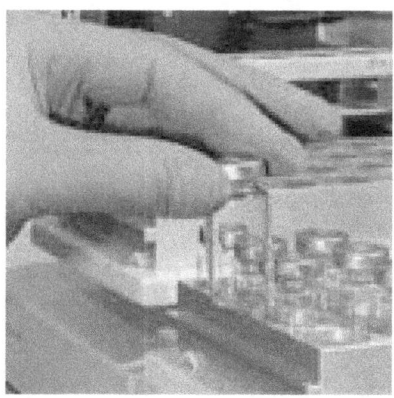

Report Contributors:

John Bishop
Chris Dunlap
Dan Howard
Tiffine Johnson-Davis
Rich Jones
Andrew Lavenburg
Rick Beusse

Abbreviations

ARL	Army Research Laboratory
BOSC	Board of Scientific Counselors
DDA	Decision Document Analysis
EPA	U.S. Environmental Protection Agency
FY	Fiscal Year
GPRA	Government Performance and Results Act of 1993
LRP	Land Research Program
LTG	Long-Term Goal
MYP	Multi-Year Plan
NAS	National Academy of Sciences
NPD	National Program Director
NRC	National Research Council of the National Academies
OIG	Office of Inspector General
OMB	Office of Management and Budget
ORD	Office of Research and Development
ORMA	Office of Resource Management and Accountability
OSTP	Office of Science and Technology Policy
PART	Program Assessment Rating Tool
PBM SIG	Performance-Based Management Special Interest Group
R&D	Research and Development
SoSP	Science of Science Policy

Cover photo: EPA scientist performing a chemical assay. (EPA photo)

U.S. Environmental Protection Agency
Office of Inspector General

10-P-0176
August 4, 2010

At a Glance

Catalyst for Improving the Environment

Why We Did This Review

We conducted this evaluation to determine whether one of the U.S. Environmental Protection Agency's (EPA's) research programs – the Land Research Program (LRP) – has appropriate performance measures for assessing the effectiveness of its research products.

Background

EPA relies on sound science to safeguard human health and the environment. LRP provides the science and technology to help its clients preserve the Nation's land, restore contaminated properties, and protect public health from exposure to environmental contaminants. LRP measures research performance by using (1) Office of Management and Budget (OMB) Program Assessment Rating Tool (PART) measures, (2) client feedback, and (3) peer review by the Board of Scientific Counselors (BOSC).

For further information, contact our Office of Congressional, Public Affairs and Management at (202) 566-2391.

To view the full report, click on the following link:
www.epa.gov/oig/reports/2010/ 20100804-10-P-0176.pdf

EPA's Office of Research and Development Performance Measures Need Improvement

What We Found

The difficulty of measuring research performance has been recognized by the National Research Council of the National Academies and other authoritative sources. No single measure can adequately capture all elements of research performance. Therefore, EPA's Land Research Program (LRP) has employed a variety of methods to assess its research performance. We found that improvements were needed to better enable EPA's Office of Research and Development (ORD) to assess the effectiveness of LRP research products. LRP did not have measures that assessed progress towards short-term outcomes identified in the LRP Multi-Year Plan (MYP). Additionally, LRP's citation analysis PART measures were not meaningful to ORD program managers and were not linked to LRP's goals and objectives. As implemented, ORD's survey of LRP clients did not provide a meaningful measure of customer feedback because ORD's client survey was not reliable. Further, LRP lacks some key measures that would aid BOSC in conducting its LRP program reviews, and ORD has not clearly defined elements of its long-term goal (LTG) rating guidance for BOSC reviews.

Several underlying issues impacted ORD's development of LRP performance measures. These include the inherently difficult nature of establishing outcome-oriented research measures and ORD's decision not to tailor its measures to each research program. As a result, ORD has invested resources in performance measures and tools that have not effectively measured key aspects of LRP performance. The measures have not provided LRP with the data to assess program progress towards its goals, identify areas for program improvement, and track the short-term outcomes of its research.

What We Recommend

We made a number of recommendations to ORD to improve LRP's research measures, including that ORD (1) develop measures linked to the short-term outcomes in LRP's MYP, (2) augment LRP's citation analysis with measures meaningful to ORD program managers and linked to LRP's goals and objectives, (3) develop an implementation plan for the LRP client survey to ensure that LRP has a reliable method for assessing relevance (or develop a reliable alternative customer feedback mechanism), (4) provide appropriate performance measurement data to BOSC prior to full program reviews, and (5) revise its LTG rating guidance to BOSC for program reviews. ORD generally agreed with our recommendations and is taking action to implement four recommendations. However, for three recommendations closely linked to the OMB PART, ORD is awaiting additional guidance from OMB before proposing specific corrective actions. We consider these three recommendations open and unresolved.

UNITED STATES ENVIRONMENTAL PROTECTION AGENCY
WASHINGTON, D.C. 20460

August 4, 2010

<u>MEMORANDUM</u>

SUBJECT: EPA's Office of Research and Development Performance Measures
Need Improvement
Report No. 10-P-0176

FROM: Arthur A. Elkins, Jr.
Inspector General

TO: Paul Anastas
Assistant Administrator for Research and Development

This is our report on the subject evaluation conducted by the Office of Inspector General (OIG) of the U.S. Environmental Protection Agency (EPA). This report contains findings that describe the problems the OIG has identified and corrective actions the OIG recommends. This report represents the opinion of the OIG and does not necessarily represent the final EPA position. Final determinations on matters in this report will be made by EPA managers in accordance with established audit resolution procedures.

The estimated cost of this report – calculated by multiplying the project's staff days by the applicable daily full cost billing rates in effect at the time – is $1,149,970.

Action Required

In accordance with EPA Manual 2750, you are required to provide a written response to this report within 90 calendar days. Your response will be posted on the OIG's public Website, along with our comments on your response. Your response should be provided in an Adobe PDF file that complies with the accessibility requirements of section 508 of the Rehabilitation Act of 1973, as amended. If your response contains data that you do not want to be released to the public, you should identify the data for redaction. For recommendations 2-3 and 2-5, the Office of Research and Development (ORD) submitted a corrective action plan that sufficiently addressed these recommendations. For recommendations 2-1 and 2-4, ORD provided subsequent written clarification on July 19, 2010 that, together with its corrective action plan, sufficiently addressed these recommendations. Accordingly, we are "closing" these four recommendations in our tracking system upon issuance of this report. No further response is required for these four recommendations. These recommendations will be tracked to completion in the Agency's tracking system. For the other recommendations, the Agency's

90-day response should include a corrective action plan for agreed-upon actions, including milestone dates. You may also propose alternative actions you believe will meet the intent of our recommendations. We have no objections to the further release of this report to the public. This report will be available at http://www.epa.gov/oig.

If you or your staff have any questions regarding this report, please contact Wade Najjum, Assistant Inspector General for Program Evaluation, at (202) 566-0832 or najjum.wade@epa.gov; or Rick Beusse, Director for Program Evaluation – Air & Research Issues, at (919) 541-5747 or beusse.rick@epa.gov.

Table of Contents

Chapters

Appendices

Chapter 1
Introduction

Purpose

The U.S. Environmental Protection Agency (EPA) relies on sound science to safeguard human health and the environment. As the scientific research arm of the Agency, EPA's Office of Research and Development (ORD) conducts research on ways to prevent pollution, protect human health, and reduce risk. Twelve ORD national research programs provide the science to support EPA's goals in its strategic plan. We conducted this evaluation to determine whether one of ORD's research programs, the Land Research Program (LRP), has appropriate performance measures for assessing the effectiveness of its research products.

Background

ORD's Land Research Program

The LRP provides the science and technology to preserve the Nation's land, restore contaminated properties, and protect public health from exposure to environmental contaminants. The science and technology developed by the LRP are used by EPA, States, local communities, regulated and responsible parties, and contractors to assess, minimize, and manage the risks of hazardous waste contamination. For example, in 2007 LRP conducted research on dredging residuals[1] from dredging operations to help develop a reliable method for estimating the volume and contaminant concentration of such dredging residuals, a high research priority for EPA's Superfund Program. Through its research, the LRP supports EPA's Office of Solid Waste and Emergency Response. The LRP Multi-Year Plan (MYP) identified two long-term goals (LTG):

- **LTG 1:** "Clients request and apply ORD research products and services needed for mitigation, management, and long-term stewardship of contaminated sites."
- **LTG 2:** "Clients request and apply ORD research products and services needed to manage material streams, address emerging material streams, and conserve resources."

---

[1] Post-dredging residuals are materials not captured by the dredge, sediments adjacent to the dredge cut that fall into the dredge footprint, suspended sediments that settle into the footprint, and suspended sediments transported downstream outside of the dredge footprint.

The LRP spent $186.2 million on land-related research in the last 5 fiscal years (FYs) (FY 2005 – FY 2009 enacted budget). Its FY 2010 resources include $36.4 million and 155 staff years (Full-Time Equivalents[2]).

Measuring Federal Research Performance

The difficulty of measuring research performance has been recognized by the National Research Council of the National Academies (NRC) and other authoritative sources. According to the NRC, agencies and oversight bodies are challenged in evaluating research programs. No single measure can adequately capture all elements of research performance. Therefore, research programs are often evaluated by assessing performance in multiple areas by using a variety of methods.

In its 2008 Guide to the Program Assessment Rating Tool (PART),[3] the Office of Management and Budget (OMB) identified relevance, performance, and quality criteria that can be used to assess the effectiveness of federal research and development (R&D) programs. Other reliable sources have also identified key attributes similar to those presented in the 2008 OMB Guide to the PART as criteria for measuring federal research performance. For example, a 2008 NRC report[4] stated that research program efficiency must be evaluated in the context of relevance, effectiveness, and quality. Similarly, a 1997 report by the U.S. Army Research Laboratory's (ARL) Special Projects Office[5] identified program relevance, productivity, and quality as the most important attributes of its research program's performance.

Research organizations can use several tools to assess key performance attributes of federal research such as relevance, productivity, and quality. For example:

- The NRC recommended that EPA and other agencies use expert review panels to evaluate the investment efficiency[6] of the relevance, performance, and quality of the research. According to the NRC report, EPA is providing its expert review committees with two kinds of data:

---

[2] A figure calculated from the number of full-time and part-time employees in an organization that represents these workers as a comparable number of full-time employees.

[3] OMB's Program Assessment Rating Tool Guidance No. 2008-01, dated January 29, 2008.

[4] *Evaluating Research Efficiency in the U.S. Environmental Protection Agency*, Committee on Evaluating the Efficiency of Research and Development Programs at the U.S. Environmental Protection Agency, Committee on Science, Engineering, and Public Policy, Policy and Global Affairs, Board on Environmental Studies and Toxicology, Division on Earth and Life Studies, National Research Council, National Academies, 2008.

[5] *Measuring Performance at the Army Research Laboratory: The Performance Evaluation Construct*, Edward A. Brown, U.S. Army Research Laboratory's Special Projects Office, published in the Journal of Technology Transfer, Vol. 22 (2): 21-26 (June 1997).

[6] The NRC report stated that *investment efficiency* concerns three questions: (1) are the right investments being made; (2) is the research being performed at a high level of quality; and (3) are timely and effective adjustments made in the multi-year course of the work to reflect new scientific information, new methods, and altered priorities? Investment efficiency is determined by examining a program in light of its relevance, performance, and quality.

(1) pilot surveys that evaluate how the research is being used, and
(2) bibliometric analyses.[7]

- ARL's 1997 report identified peer review, customer evaluations, and metrics as evaluation tools that can be used to assess program relevance, productivity, and quality. The report stated that if peer review was independent and of "sufficient stature" it would answer the question concerning quality, and that customer evaluation would address the issues of relevance and productivity. ARL considered metrics to be an adjunct to the two other methods.

Consistent with the measurement methods and tools recommended by these groups, the LRP assesses the effectiveness of its federal research and development program by using a variety of methods, including:

- Quantitative measures that were developed for the OMB PART, including bibliometric analysis;

- EPA/ORD conducted customer feedback mechanisms; and

- Board of Scientific Counselors (BOSC) external peer reviews.[8]

Details on each of these measures follow.

Quantitative Measures Developed For OMB PART Review

Performance measures are key factors that OMB considers in its PART program ratings. Because a program's performance goals represent its definition of success, the quality of the performance goals and actual performance against those goals are the primary determinants of an overall PART rating.[9] The LRP has four output measures and one efficiency measure.

PART Output Measures:

1. "Percentage of Land research publications rated as highly cited publications."
2. "Percentage of Land publications in "high impact" journals."[10]

---

[7] According to ORD guidance, bibliometrics is the study or measure of published information. The phrase "bibliometric analysis" is an umbrella phrase which includes various types of analyses. Some of these analyses types will be defined and discussed later in this report. Bibliometric analyses are used to explore the impact of a particular paper, set of papers in a field, or set of researchers.
[8] BOSC provides qualitative assessments of charge questions relating to the relevance, quality, performance, and scientific leadership of the program.
[9] See footnote 3.
[10] According to PART, high impact journals are an indication of quality and influence. This measure evaluates the percentage of LRP publications that are accepted in prestigious journals and their subsequent impact on the field. The criteria and "impact factor" rankings for this metric are provided by Thomson's Journal Citation Reports.

3. "Percentage of planned outputs delivered in support of the mitigation, management and long-term stewardship of contaminated sites long-term goal."
4. "Percentage of planned outputs delivered in support of the manage material streams, conserve resources and appropriately manage waste long-term goal."

PART Efficiency Measure:

5. "Average time (in days) for technical support centers to process and respond to requests for technical document review, statistical analysis and evaluation of characterization and treatability study plans."

The LRP was most recently assessed under OMB's PART in 2006. OMB rated the program as "Adequate." ORD was using the above PART measures at the time of our evaluation although the PART process was under review by OMB. Information for the first two PART measures identified above is obtained through *citation analysis*, a type of bibliometric analysis. *Citation analysis* identifies each time selected ORD publications are cited in peer-reviewed scientific literature.

During 2009, EPA issued a work assignment to an existing contract that, when amended, committed 2,900 hours and about $222,000 to obtain bibliometric analyses between April 1, 2009, and March 31, 2010. This work assignment, which may be used by all ORD labs and centers, provides for the contractor to perform a type of bibliometric analysis called *decision document analysis* (DDA). This type of analysis identifies each time selected ORD publications are cited in EPA, other federal, State, or comparable international organization's decision documents. Decision documents include regulations, records of decision, and policy guidelines. According to an ORD program analyst, the DDA concept was developed by ORD and the first DDA was completed in 2008 for the LRP. ORD's Office of Resource Management and Accountability (ORMA) estimated the contract cost for this analysis was $8,000 and the accompanying citation analysis was $15,600. ORD is evaluating the benefits of expanding DDA to use in program assessment. This evaluation includes a pilot study. ORD plans to obtain feedback about the usefulness of research products from the people/organizations that requested the research. An ORD document summarized the ongoing effort as follows:

> *DDA are currently in the trial phase, but are expected to be a key indicator of program performance—specifically, the data should help indicate the relative quality, relevance, and impact of ORD's science.*

The contract also allows an EPA work assignment manager to request a determination of how ORD research publications were used. Specifically, the contract states: "As a part of this analysis the contractor shall determine whether the research provided special tools or methods, baseline characterization, dose response relationships, etc."

LRP's Survey of Customers

From March to April 2008, ORD conducted a survey, known as a "partner survey," of the LRP's customers/clients in EPA. The survey was to obtain EPA client feedback on LRP's effectiveness and the extent to which LRP's clients used its science products and technical support in their decision making. This survey was to serve as a program management and performance assessment tool.

The survey results stated that 70 percent of the respondents were satisfied with the quality of LRP products and services. The survey also stated that about 70 percent of the program offices and regions were "less than mostly satisfied" with LRP's (1) communications with clients on research project progress; (2) willingness to involve clients in research planning; (3) lack of flexibility in accommodating client needs; and (4) on-time delivery of products and services.

BOSC External Peer Reviews

The BOSC is a federal advisory committee established by EPA in 1996 to provide advice and recommendations to ORD. EPA first began using BOSC to conduct external peer reviews of its research programs in 2004, after considering other peer review options, including contractors. The BOSC consists of an Executive Committee (approximately 15 members) and smaller subcommittees that are created, as needed, to evaluate individual ORD labs, centers, and research programs. In selecting BOSC members, EPA considers candidates from the environmental scientific/technical fields, health care professionals, academia, industry, public and private research institutes or organizations, and other relevant interest areas. According to the BOSC Executive Committee Designated Federal Officer, ORD's Deputy Assistant Administrator for Science wanted the process to be as transparent as possible and believed that BOSC, since it is subject to the Federal Advisory Committee Act,[11] would provide an independent and transparent process.

The draft Board of Scientific Counselors Handbook for Subcommittee Chairs states that BOSC subcommittees should evaluate the quality, relevance, structure, performance, scientific leadership, and coordination/communication of ORD research programs every 4-5 years during full program reviews. These subcommittees also conduct mid-cycle reviews to review ORD's progress in addressing BOSC recommendations from prior full program reviews. The estimated annual operating cost of the BOSC is $1,130,500, which includes 4.3 person-years of support. BOSC conducts from six to eight EPA ORD reviews annually. ORD costs for these BOSC reviews range between $500,000 to $700,000 per year. BOSC's 2006 full review report on LRP stated that the program is meeting its stated goals. The report also stated that the Land MYP

---

[11] The Federal Advisory Committee Act became law in 1972 and is the legal foundation defining how federal advisory committees operate. The law has special emphasis on open meetings, chartering, public involvement, and reporting.

provided an adequate roadmap for achieving the LRP's multi-year goals in the future. BOSC's 2009 mid-cycle review report stated that the program exceeds expectations, and was responsive to the recommendations put forth in the 2006 full BOSC review report.

Noteworthy Achievements

EPA's FY 2009 Performance and Accountability Report listed measuring the results of research as a 2010 challenge not only for EPA, but also for the broader research community. To improve the measures for assessing the effectiveness of federal research and development programs, ORD has been an active participant and leader in the Science of Science Policy (SoSP) community, an interagency effort fostered by the U.S. Office of Science and Technology Policy (OSTP).[12] In December 2008, ORD personnel attended the first SoSP workshop.[13] This began a dialogue between academic researchers and federal practitioners, to better understand the efficacy and impact of Federal Research and Development investment. In October 2009, EPA ORD and the U.S. Department of Agriculture together led a second SoSP workshop, titled *What's in YOUR Tool Box? Best Practices in R&D Prioritization, Management, and Evaluation,* an interagency effort to improve assessing research effectiveness. The workshop included presentations and discussions during which R&D agencies presented their best practices in Federal R&D prioritization, management, and evaluation, and how metrics improve Federal R&D efficiency.

Scope and Methodology

We conducted our evaluation from May 2009 to May 2010, to focus on the performance measures and tools used by ORD's LRP. However, the issues we identified, and our recommendations to address those issues, may be applied to some or all of ORD's 11 other research programs. We conducted this evaluation in accordance with generally accepted government auditing standards. Those standards require that we plan and perform the evaluation to obtain sufficient, appropriate evidence to provide a reasonable basis for our findings and conclusions based on our evaluation objectives. We believe the evidence obtained provides a reasonable basis for our findings and conclusions based on our evaluation objectives. Appendix A provides details on our scope and methodology.

---

[12] OSTP was established by Congress in 1976 to advise the President and others within the Executive Office of the President on the effects of science and technology on domestic and international affairs, and to lead interagency efforts to develop and implement sound science and technology policies and budgets.

[13] *The Science of Science Policy: A Federal Research Roadmap.*

Chapter 2
ORD Performance Measures Need Improvement

The Land Research Program has employed a variety of methods to assess its research performance. These include (1) OMB PART measures, (2) customer/client feedback, and (3) peer review by the BOSC. We found that improvements are needed in all three areas to better enable ORD to assess LRP's effectiveness. Specifically:

- some of the LRP's PART measures were not appropriate for use in assessing program performance; and

- as implemented, ORD's survey of LRP clients did not provide a meaningful measure of customer feedback.

Further, the following two issues limit the usefulness of BOSC reviews:

- the LRP lacks some key measures to aid in BOSC program reviews; and

- ORD has not clearly defined elements of its long-term goal rating guidance for BOSC program reviews.

Several underlying issues impacted ORD's development of LRP performance measures. These include the inherently difficult nature of establishing outcome-oriented research measures and ORD's decision not to tailor its measures to each research program, but rather to use similar measures across all of its research programs for PART. Additionally, according to the ORMA Accountability Team Leader, the matrix management approach employed in ORD's organizational structure also impacted ORD's development of LRP performance measures. As a result, ORD has invested resources in performance measures and tools that have not effectively measured key aspects of LRP performance. The measures have not provided LRP with the data to assess program progress towards its goals, identify areas for program improvement, or track the short-term outcomes of its research.

Some LRP PART Measures Not Appropriate to Assess Program Performance

Not all of the measures included in the LRP's suite of PART measures were appropriate to assess the research program's performance. For example, the LRP did not have measures that assessed progress towards short-term outcomes identified in the LRP Multi-Year Plan (MYP). Additionally, the LRP's citation analysis PART measures were not meaningful to ORD program managers and were not linked to the program's goals and objectives. Appendix B lists these and other criteria for evaluating performance measures.

Some of the LRP's individual PART measures, however, may provide benefit to the program when considered as part of a larger suite of effective performance measures. For example, LRP's MYP provides annual performance goals and annual performance measures[14] that are used to track research outputs. Within the framework of its MYP, the LRP measured whether its research outputs for each LTG were completed in a timely manner. The LRP also measured the efficiency of its technical support centers by measuring how long these centers took to respond to technical requests made by EPA regions and program offices.

Although the LRP has taken steps to improve its measurement of program outcomes and impacts, it continues to use its existing PART measures. ORD anticipates that OMB will revise its PART guidance and requirements, and has been waiting on new OMB guidance before revising its existing LRP measures. However, ORD included the existing LRP PART measures in its Fiscal Year 2011 Annual Performance Plan and Congressional Justification. ORD plans to continue to use its PART measures, even if the PART process is abandoned or revised, because these measures are required under the Government Performance and Results Act of 1993 (GPRA), according to the ORMA Accountability Team Leader. Regardless of how LRP's existing PART measures are reported in the future, some of the LRP's existing PART measures will not allow LRP to assess progress towards its goals and objectives.

LRP PART Measures Do Not Assess Short-Term Outcomes

Despite OMB requirements for federal programs to use outcome-oriented performance measures, none of the LRP's five PART measures assess the short-term outcomes[15] identified in the LRP MYP. Outcome-oriented performance measures allow for an assessment of the results of a program activity as compared to its intended purpose. Such measures are required by OMB unless the program can justify why it is unable to define satisfactory quantifiable outcome measures. OMB's PART guidance does not distinguish outcomes measures as long-term, intermediate, or short-term. A 2008 National Research Council of the National

---

[14] According to ORD's April 2008 guidance for preparing MYPs, the MYP structure forms the foundation for ORD's performance measurement, including annual performance tracking of key activities and outputs. MYPs outline the sequence and relationship of annual performance goals needed to achieve each long-term goal (LTG), and the annual performance measures that contribute to each annual performance goal. ORD defines annual performance goals as major research outputs that are described in the context of the outcome to which they contribute. They represent *significant* milestones along a critical path toward accomplishing the LTGs. Annual performance measures are defined as research outputs that contribute to accomplishing an annual performance goal by addressing the most important scientific issues for that particular annual performance goal. Annual performance measures may include research products such as reports, journal articles, models, and evaluations.

[15] The short-term outcomes identified in the Land Research Program's Multi-Year Plan (Fiscal Years 2007–2012) are related to the utilization of research products by the Office of Solid Waste and Emergency Response and EPA regions.

Academies (NRC) report[16] examined the use of outcome measures by research organizations in detail, and stated that it was not feasible for ORD to establish measures for its ultimate (i.e., long-term) outcomes. The NRC report stated that ultimate outcome-based metrics for evaluating the efficiency of research are neither achievable nor valid because:

- there is often a large gap in time between completing research and the ultimate outcome of the research;
- a number of entities over which the research program has no control is responsible for translating research results into outcomes; and
- the results of research may change the nature of the outcome.

However, while it may not be possible to establish measures for ultimate outcomes, the 2008 NRC report said that federal research agencies should measure intermediate and/or short-term outcomes.

The short-term outcomes that are expected from LRP research activities include client use of research products in:

- guidance
- records of decision
- site-specific applications
- regulations
- risk management decisions

Although the LRP identified short-term outcomes in its MYP, it did not establish any PART performance measures that were linked to these short-term outcomes. OMB identified the LRP's PART measures as either output or efficiency measures. Figure 2.1 provides OMB's definition of efficiency, outcome, and output performance measures. Output and efficiency measures can be useful for managing program performance; however, these types of measures do not provide an assessment of progress towards the intended outcomes of the program.

---

[16] *Evaluating Research Efficiency in the U.S. Environmental Protection Agency*, Committee on Evaluating the Efficiency of Research and Development Programs at the U.S. Environmental Protection Agency, The National Academies. ISBN: 0-309-11685-6. The National Academies Press, Washington, DC, 2008.

Figure 2.1: OMB Definitions for Selected Types of Performance Measures

OMB Definitions for Efficiency, Outcome, and Output Performance Measures
Efficiency measure: Efficiency measures reflect the economical and effective acquisition, utilization, and management of resources to achieve program outcomes or produce program outputs. **Outcome measure:** A measure of the intended result or impact of carrying out a program or activity. They define an event or condition that is external to the program or activity and that is of direct importance to the intended beneficiaries and/or the general public. **Output measure:** Outputs describe the level of activity that will be provided over a period of time, including a description of the characteristics (e.g., timeliness) established as standards for the activity. Outputs refer to the internal activities of a program (i.e., the products and services delivered).

Source: *Guide to the Program Assessment Rating Tool (PART)*. Program Assessment Rating Tool Guidance No. 2008-01. Office of Management and Budget, January 2008.

Although the LRP has not revised its existing PART measures, ORD personnel have recently started using the following performance measurement tools to help measure short-term outcomes:

- Decision Document Analysis (DDA) - identifies each time selected ORD publications are cited in decision documents of EPA or other federal, State, or comparable international organizations. Decision documents include regulations, records of decision, and policy guidelines. ORD is evaluating the benefits of expanding DDA for use in program assessment. See Chapter 1 for a more detailed description.

- Site-specific summary tables – In 2009, the LRP National Program Director (NPD) began tracking site-specific short-term outcomes for ground water research. The NPD said that tracking the users of LRP research is important and that this effort is more useful than tracking the current PART measures.

Without short-term outcome measures, it is difficult for the LRP to assess whether its work is meeting its intended purpose and goals.

Current Citation Analysis Measures Do Not Meet Key Criteria for Effective Performance Measurement

The LRP citation analysis measures (PART measures 1 and 2) are not meaningful to ORD program managers, and are not linked to program goals and objectives. Based on our review of performance measurement literature, we concluded that such elements were key characteristics of effective performance measures. See Appendix B for an OIG-developed list of criteria for evaluating performance measures.

Meaningful to ORD Program Managers

The citation analysis measures employed by the LRP are not meaningful to the ORD program managers, particularly the NPD. In our opinion, a measure should demonstrate several characteristics to be considered meaningful to internal stakeholders, such as ORD program managers.[17] However, the LRP's citation analysis measures do not meet key characteristics we identified for effective measures. For example, the measures do not drive actions required to achieve strategic goals because the measures are not linked to the LRP's goals, objectives, or short-term outcomes (this is discussed in further detail below).

Further, the LRP's citation analysis measures are not used by the program managers, or perceived by some ORD personnel as valuable tools to managing the program. According to ORMA, the citation analysis PART measures are not widely used by ORD program managers. The LRP's citation analysis measures were adopted by ORD in an effort to satisfy OMB's PART requirements, according to ORMA's Acting Accountability Team Leader.

Linked to Program Goals and Objectives

The LRP's citation analysis measures are not linked to the program's long-term goals, or the short-term outcomes identified in its MYP. The LRP's long-term goals are related to clients' requests and use of its research products. However, the citation analysis measures do not identify information regarding the use of the LRP's research products by its primary clients: EPA program and regional offices. Therefore, these measures do not indicate the LRP's progress towards program goals or its strategic objectives.

According to ORD, the LRP's citation analysis measures (i.e., percentage of highly cited publications, and percentage of publications in high-impact journals) are indicators of research quality. However, existing literature presents multiple opinions regarding what citations of scientific work actually measure. For example, a 2005 book on citation analysis in research evaluation stated that it is "… extremely difficult if not impossible to express what citations measure in a single theoretical concept…"[18] A 2009 journal article discussing the use and misuse of journal metrics and other citation indicators stated that:

> *Generally, citations represent the notions of use, reception, utility, influence, significance, and the somewhat nebulous word "impact." Citations do not, however, represent measures of quality. Quality assessments require human judgment.[19]*

---

[17] See Appendix B, number 5.

[18] Moed, Henk F., 2005, *Citation Analysis in Research Evaluation.*

[19] Pendlebury, David A., 2009, "The Use and Misuse of Journal Metrics and Other Citation Indicators." *Archivum Immunologiae et Therapiae Experimentalis* 57(1):1-11.

Several studies have shown a correlation between highly cited publications and other indicators of research quality (e.g., peer review results). Therefore, when used as a supplementary tool in a peer review process, citation analysis measures can aid in assessments of research quality. However, in an August 2009 meeting, one member of BOSC's Executive Committee, the group charged with conducting peer review evaluations of ORD research program performance, stated that ORD's bibliometric analysis measures (i.e., citation analysis measures) were not effective in communicating the quality or quantity of research that had been accomplished. Another member stated that the citation analysis measures did not provide information to address the full scope of BOSC reviews.

For citation analysis to be an effective indicator of quality, it should be measuring aspects of program performance that are linked to the LRP's definition of research quality, as well as its goals and objectives for research quality. The term "quality" can be defined differently by different programs and stakeholders. For example, the National Performance Review, in its 1997 Benchmarking Study Report, discusses the quality of an output in terms of the extent to which a client is satisfied with the output. In that sense, citation analysis would not provide a measure of quality for the LRP.

ORD's 2001 Strategic Plan states:

> *EPA is a mission-oriented agency; its work is focused on the specific goals and objectives described in EPA's Strategic Plan. ORD provides the scientific support that enables the Agency to meet those goals and objectives. EPA depends on ORD; its scientific reputation is built on our work. As a result, our research must respond to and anticipate the Agency's needs. Science that is not supportive of EPA's mission may be important, but it need not, and should not, be done by ORD.*

Measures that are not meaningful to internal stakeholders (such as ORD program managers), and not linked to the program's long-term goals, can have negative effects on the research program. For example, a 1999 National Academy of Sciences (NAS) report[20] stated that for measures such as citation analysis, "...researchers are apt to take actions that artificially increase the number of citations they receive..." The NAS also found that researchers are apt to "... reduce their research in fields that offer less opportunity of immediate or frequent publication or in critical related fields ... that do not offer publication opportunities." Such actions can create goal displacement within the research program, and result in missed opportunities to focus research on EPA's priority needs.

---

[20] *Evaluating Federal Research Programs: Research and the Government Performance and Results Act*, Committee on Science, Engineering, and Public Policy (COSEPUP), National Academy of Sciences, National Academy of Engineering, Institute of Medicine, ISBN: 0-309-51798-2. National Academy Press, Washington, DC, 1999.

LRP PART Measures Not Effective Because ORD Did Not Establish Program-Specific PART Measures

Some of the LRP PART measures were not effective in assessing the progress towards LRP goals and objectives because ORD decided to use similar measures across all of its research programs. ORD did not create measures to address the specific goals and objectives of LRP. Factors that contributed to this approach included the difficult nature of establishing outcome-oriented measures of research products, OMB's PART requirements, and the organizational structure employed by ORD.

Research outcomes are generally difficult to track. The NRC found that the useful outcomes of basic research cannot be measured on an annual basis because their timing and nature are inherently unpredictable.[21] OMB acknowledged this point in a 2004 budget statement regarding research, noting that "It is preferable to have outcome measures, but such measures are often not very practical to collect or use on an annual basis." The LRP NPD also stated that no automatic outcome occurs from a given technical report, and that it sometimes takes years for the research product to be applied or used. This time lag makes it difficult for the LRP to get accurate client feedback on using some of its more recent research products.

According to the ORMA Accountability Team Leader, ORD tried to come up with a standard way of assessing performance across its programs that OMB would agree to. The resulting measures, according to ORMA's Acting Accountability Team Leader, were put in place to comply with OMB PART requirements. ORD only recently made advances in developing measures of its short-term outcomes (see information above related to DDA and site-specific summary tables).

Although each of ORD's research programs has unique research outputs, long-term goals, and expected outcomes, ORD applied similar PART measures to all of its research programs in an effort to establish a general performance measurement framework to propose to OMB. As part of this framework, the NPDs were responsible for working with ORD's labs, centers, and offices to get agreement on the research outputs for a given research program.

ORMA's Accountability Team Leader stated that there is no clear line of accountability for who is responsible for managing performance in ORD's research programs. He said that each level of ORD's organization helps manage research performance. The NPDs are held accountable for research program performance, yet it is the lab/center directors that have the ultimate responsibility to develop ORD's research products and decide how they are produced. According to ORMA's Accountability Team Leader, if ORD research programs adopted a variety of performance measures, it would be difficult for lab/center

---

[21] See footnote 16.

directors to track the measures for each research program. Due in part to its management structure, ORD adopted a one-size-fits-all performance measurement approach for PART. ORD stated in its response to our draft report that the matrix management structure has assisted ORD in managing the complex nature of the research being conducted. However, in regard to performance measurement, we found that the matrix management structure contributed to LRP PART measures that are not effective for assessing the progress towards the LRP's goals and short-term outcomes.

ORD ORMA is waiting on new guidance from OMB before revising any PART measures, and continues to use the current measures. EPA included the existing LRP PART measures in its Fiscal Year 2011 Annual Performance Plan and Congressional Justification.

ORD's Survey of LRP Clients Did Not Provide a Meaningful Measure of Customer Feedback

ORD's 2008 survey of LRP clients did not provide a meaningful measure of customer feedback because ORD's client survey was not reliable. According to ORD, the LRP survey was the first survey of its kind and results of the process were used to develop ORD guidance for subsequent surveys.

In designing and implementing the client survey, ORD did not:

- identify the universe of LRP clients before administering the survey;
- conduct a representative sample of LRP clients; or
- obtain a sufficient number of responses (and did not adhere to its own criteria for minimum number of responses).

As a result, the survey was not a reliable measure of LRP performance. The *ORD Partner Survey Methodology* document stated that:

> *The partner list should be (1) inclusive of all program partners or (2) a representative sample large enough to ensure a high confidence level of results. The ideal sample size to maintain a 95% confidence level +/-10% should be no less than 100 respondents. If 100 partners cannot be identified to respond to the survey, then the sample should include as many partners as possible, knowing that the confidence level will diminish as the number of respondents decreases.*

However, ORD did not develop a complete universe of LRP's clients. According to ORMA, the clients that received the survey were recommended by the LRP NPD with feedback from the Research Coordination Team.[22] Those recommendations resulted in the survey being sent to 103 EPA clients in EPA

---

[22] The Land Research Coordination Team was established by EPA ORD to facilitate research planning and communication in the Land Research Program so that research meets the highest priority needs of the Agency.

regions and program offices. Of the 103 surveys distributed, ORD received 50 responses. This was not a representative survey of LRP's clients because it did not include all clients, was not large enough to ensure confidence in the results, did not obtain the minimum number of 100 respondents desired, and did not sample a sufficient number of clients to ensure statistical validity.

ORD's draft 2009 guidance states that *Client Survey Data* and *Client Use of ORD Research* are two pieces of evidence that BOSC should use in assessing the quality and impact of the research program. Similarly, ARL[23] listed three main evaluation tools (i.e., peer review, metrics, and customer evaluation) for assessing performance at research organizations. ARL stated that customer evaluation was very useful in assessing research relevance and productivity.

In 2009, BOSC also identified concerns with ORD's survey, noting that ORD's client survey should be improved so that it is informative or it should be abandoned. As currently conducted, BOSC found ORD's client surveys of limited utility. As of April 2010, ORMA was reassessing its approach to client surveys and looking into other ways of obtaining customer feedback.

The current status of the LRP survey is unclear. While we understand that ORMA is reassessing its approach to surveys and looking into other potential avenues for collecting customer feedback, the LRP must have a reliable tool for gathering performance data. Without a reliable method to assess the relevance of its research products, the LRP lacks sufficient information to determine:

- how well the organization is doing,
- the organization's progress toward meeting its goals,
- customers' satisfaction with research produced, and
- areas of necessary improvement.

LRP Lacks Some Key Measures to Aid BOSC Program Evaluations

Key performance measures were not in place for the 2006 BOSC review of the LRP, and current LRP measures are not sufficient to aid in BOSC program evaluations. Since 2006, ORD has taken steps to improve its guidance for BOSC program reviews, such as including a list of performance rating materials that research programs should supply to BOSC prior to program reviews. Several underlying issues impacted ORD's development of LRP performance measures. These include the inherently difficult nature of establishing outcome-oriented research measures and ORD's decision not to tailor its measures to each research program, but rather to use similar measures across all of its research programs for PART. Additionally, according to the ORMA Accountability Team Leader, the matrix management approach employed in ORD's organizational structure also impacted ORD's development of LRP performance measures. As a result, BOSC may not be able to effectively recommend actions to improve research program

---

[23] See footnote 5.

quality, relevance, structure, performance, scientific leadership, coordination/communication, and outcomes.

Key Performance Measures Not in Place for 2006 BOSC Review

At the time of the 2006 BOSC review, the LRP did not have measures in place to capture client feedback data or to measure the short-term outcomes of its research products. These measures were not in place because ORD had not fully developed performance measurement tools such as client surveys and decision document analysis by the time of the 2006 BOSC review. Consequently, BOSC was left to draw conclusions for certain aspects of its 2006 review without comprehensive performance data on the relevance of LRP's products, and its progress towards reaching short-term outcomes.

In its 2006 review, BOSC was tasked by ORD with answering five program assessment charge questions. ORD asked BOSC to consider several factors in evaluating each charge question. Table 2.1 presents selected charge questions and "factors considered"[24] from the 2006 BOSC review aimed at determining whether the LRP met stakeholder[25] needs and whether LRP products were used by clients and stakeholders.

Table 2.1: Selected Charge Questions and Factors Considered from BOSC's 2006 Review of ORD's Land Research Program

Review Category	Charge Question	Selected Factors Considered
Relevance	*Is the research program relevant to and consistent with Agency goals and customer needs, and is it sufficiently flexible?*	• Is the program responsive to Agency and other stakeholder needs and priorities? • Are outputs from the program used by stakeholders?
Performance	*Is the research program making timely progress in addressing key scientific questions and LTGs?*	• Has the program made significant progress toward each of the LTGs? • Has the program met stakeholder needs in a timely and useful way? • Is there evidence for application of products and knowledge by clients that would lead toward achieving program outcomes?

Source: Selected information from BOSC's 2006 final report, *Review of the Office of Research and Development's Land Restoration and Preservation Research Program at the U.S. Environmental Protection Agency.*

---

[24] BOSC was tasked with answering five charge questions addressing four review categories (relevance, quality, performance, and scientific leadership) in its 2006 review of LRP. For each charge question, BOSC was tasked with addressing a series of sub-questions and issues which it refers to in its report as "factors considered." Each of the "factors considered" is addressed individually by BOSC in its report, and cumulatively addresses the overall charge question.

[25] In its 2006 review of LRP, BOSC referred to EPA program offices and regions, federal agencies, States, and responsible parties and regulated entities as stakeholders that used LRP outputs.

However, none of the corresponding sections in the 2006 BOSC report presented data to support the conclusions made. For example, BOSC concluded that the outputs generated by the LRP were used by other EPA program offices and regions, federal agencies, States, and responsible and regulated parties, but did not provide data to support how extensively these outputs were used. Further, the BOSC report stated that it was unclear how ORD intended to measure or track client use, and incorporate such data into statements about the performance and impact of the LRP. Without such measures of client use, BOSC was left to draw conclusions regarding research relevance and outcomes based only on the testimony that it obtained from LRP personnel and selected LRP clients during the program review.

As shown in Table 2.1, another factor that BOSC considered as part of its 2006 review was whether there was any "...evidence for application of the LRP's products by clients that would lead to achieving program outcomes." However, BOSC did not present quantifiable data or evidence of client use of the LRP's products to address that "factor considered." A performance measurement tool such as DDA could have provided this type of performance data to aid BOSC in addressing this factor; however, the LRP did not have a performance measure to track this type of data at the time of the 2006 BOSC review.

Key LRP Measures Not Sufficient to Aid BOSC Reviews

Since BOSC's 2006 peer review of the LRP, ORD has updated its guidance for BOSC reviews, and included a table of suggested background information and performance rating materials that should be provided to BOSC prior to program reviews. The updated guidance identifies research program annual performance goals and measures, results of bibliometric analysis, client surveys, and client testimonials as performance data that should be supplied to BOSC for program reviews. However, an ORMA manager and a BOSC Executive Committee member both told us that the performance measurement data that ORD provided to BOSC for its program reviews have not been sufficient to fully address all of BOSC's program assessment charge questions.

Independent, expert review is better suited for judging research quality than judging research relevance and performance. According to a BOSC Executive Committee member, BOSC uses a variety of methods to assess research program quality, including its own expertise. However, because peer review panels are less equipped to judge research relevance and performance, it is particularly important for ORD to provide BOSC with appropriate performance data related to the relevance and performance of the research program.

The BOSC Executive Committee member told us that the performance data provided by ORD have been useful but somewhat incomplete. He said this situation can make charge question evaluation "somewhat uneven" across reviews of different ORD research programs, labs, or centers. Specifically, he said that the performance data are generally inadequate to evaluate the program from an investment efficiency perspective. The NPD for the LRP told us that information on client use and the short-term outcomes of the LRP's research are provided in poster presentations given to BOSC during the program review process. A BOSC Executive Committee member acknowledged that such testimonial data are considered by BOSC to evaluate program outcomes. However, he also said that the data regarding client use of research were "somewhat inconsistent" and that a tracking tool that tracks how clients use research products would be very helpful to the BOSC committees in evaluating short-term outcomes.

The program review process is highly dependent on ORD providing the right materials to fully inform the BOSC subcommittee's evaluation of the program. Therefore, reliable data must be available for BOSC to address all of its charge questions and "factors considered."

ORD Has Not Clearly Defined Elements of Its Long-Term Goal (LTG) Rating Guidance for BOSC Program Reviews

Although ORD has developed LTG ratings guidance to assist BOSC subcommittees in applying qualitative LTG ratings, aspects of this guidance have not been clearly defined. Specifically, ORD has not linked BOSC summary assessment charge questions to the LTG ratings definitions, and the LTG ratings definitions did not clearly define what constitutes program success. While the narrative summary that accompanies LTG ratings is important for managing individual programs like the LRP, the rating itself is also an indicator of ORD program performance. LTG ratings allow for cross-organization assessment, and improvements in LTG ratings indicate progress from one BOSC program review to the next, according to ORD's ORMA Accountability Team Leader. ORD guidance stated that since the BOSC qualitative rating tool is applied across all ORD research programs, it is important to apply the tool consistently. However, it has been challenging for expert review panels, including BOSC, to reach consensus on and apply qualitative program ratings. Because ORD has not clearly defined elements of its LTG rating guidance, it could not ensure that BOSC subcommittees have applied LTG ratings consistently.

ORD Has Not Linked BOSC Summary Assessment Charge Questions to the LTG Ratings Definitions

The ORD ratings methodology contained in the draft BOSC Handbook for Subcommittee Chairs states that BOSC should assess each LTG according to three summary assessment charge questions (see Figure 2.2). The summary assessment charge questions represent a subset of the program assessment charge questions, used by BOSC to evaluate the entire research program. Based on BOSC's assessment of the program's progress towards meeting LTGs, ORD asks BOSC to assign one of four qualitative ratings (i.e., Exceptional, Exceeds Expectations, Meets Expectations, and Not Satisfactory).

Figure 2.2: BOSC Summary Assessment Charge Questions

Charge Question 1
How appropriate is the science used to achieve each LTG, i.e., is the program asking the right questions, with the most appropriate methods?

Charge Question 2
How high is the scientific quality of the program's research products?

Charge Question 3
To what extent are the program results being used by environmental decision-makers to inform decisions and achieve results?

Source: *Board of Scientific Counselors Handbook for Subcommittee Chairs*, November 2008 (Draft).

However, ORD has not linked one of the three summary assessment charge questions to elements in the qualitative rating definitions. As shown in the table in Appendix C, Charge Question 3 has not been addressed by the definitions that ORD provided for each qualitative rating option. None of the LTG rating definitions provided by ORD define the extent to which program results have been used by environmental decision makers to inform decisions and achieve results. As such, it is unclear how BOSC's analysis for Charge Question 3 would affect its qualitative rating for each LTG.

Further, an element of the qualitative ratings has not been linked to any of the summary assessment charge questions. Each rating includes an assessment of the speed at which work products (tools, methods, etc.) are produced and milestones met. However, none of the summary charge questions addresses this type of assessment. If BOSC only focused on the three summary assessment charge questions, it would not be able to address the timeliness aspect of the qualitative ratings. ORD's methodology states that the qualitative LTG rating may reflect considerations beyond the summary assessment charge questions. However, the absence of a direct link between summary assessment charge questions and LTG rating definitions means that performance information considered by one BOSC subcommittee when assigning LTG ratings may not be considered by another.

The LTG Ratings Definitions Did Not Clearly Define What Constitutes Program Success

ORD guidance states that the only way to ensure consistent application of LTG ratings is for each subcommittee to follow the rating category definitions provided. BOSC subcommittees are not to interpret the rating definitions when assigning ratings. However, the ratings definitions have not clearly defined what constitutes program success. For example, ORD's definition for the "Meets Expectations" rating is as follows:

> *Meets Expectations*: indicates that the program is meeting most of its goals. Programs meet expectations in terms of addressing the appropriate scientific questions to meet their goals, and work products are being produced and milestones are being reached in a timely manner. The quality of the science being done is competent or better.

Rather than clearly defining program success as it relates to ORD's performance expectations, the above definition contains ambiguous phrases such as "most of its goals," "appropriate scientific questions," and "in a timely manner." Since the ratings definitions represent general guidance applied across programs, ORD may find it difficult to add clarity and focus to the definitions. However, ORD could supplement the definitions with program-specific milestones, and benchmarks for success, to further assist BOSC subcommittees in differentiating between the rating options.

Based on our review of performance measurement literature (see Appendix B), performance measures should be well-defined and logically designed. Well-defined measures are clear, focused, and unambiguous to avoid misinterpretation. Logically designed measures include milestones and/or indicators to express qualitative criteria, and have reliable benchmark data, standards, or alternative frames of reference for interpreting the measure. ORD guidance provides a list of performance rating materials to be considered by BOSC subcommittees in performing reviews and assigning LTG ratings. However, ORD has not established meaningful program-specific milestones for each of these items. The ORMA Accountability Team Leader stated that setting performance milestones is one of ORMA's most difficult challenges because it requires being able to measure something consistently over time. He further stated that "we haven't had enough historical trend data upon which to set truly meaningful targets--those that would be informed by past performance history or benchmarked to some ideal level of performance." Further, ORD does not provide BOSC benchmarking data that would further assist subcommittees in differentiating between the ratings options.

Without supplemental milestones and benchmarking data, ORD's ratings tool does not meet performance measurement criteria for well-defined and logically designed measures. To apply an LTG rating, BOSC members have to interpret the definitions and establish their own expectations for performance. A BOSC Land Subcommittee member expressed similar concerns, stating:

> *...most BOSC subcommittees struggle for a time to come up with meaningful and quantitative applications of the ratings. The ratings guidance provided with the charge questions are useful in that they offer perspective on why the rating is important and the types of ORD activities and accomplishment that should be considered as part of the rating. However, building a consensus opinion within the committee usually takes a while, as members first apply the guidance based on their own experience and personal perspective, and then through the subcommittee discussions there is agreement reached regarding what are reasonable "expectations" regarding the progress and quality of the science in light of the complexity and novelty of the research activity as well as funding and organizational constraints.*

As demonstrated by the Subcommittee member's statement, BOSC subcommittees may be setting their own expectations for program success, rather than evaluating programs based on consistent, well-defined performance criteria established prior to program reviews. Expectations for success should be established by ORD management, or by the directors of the research programs. However, in the absence of meaningful milestones and benchmarks, BOSC subcommittees have not had reliable reference points or standards against which program performance or achievements could be assessed. If appropriately linked to ORD's ratings definitions, such performance information would assist BOSC subcommittees in differentiating between the ratings options, and promote consistency in ratings application.

Conclusions

ORD has not developed the LRP-specific measures and tools needed to manage and improve the effectiveness of LRP research products. Individual performance measures and measurement tools within the LRP's framework are not appropriate for assessing the effectiveness of the LRP's research products or to assess progress towards short-term outcomes. Areas in each of the LRP's performance measurement tools need improvement, including the metrics provided to OMB for its PART reviews, ORD's survey of LRP clients, and the data and guidance provided to BOSC for its peer review of the LRP. The LRP needs more effective measures and tools for identifying areas for program improvement and assessing research outcomes.

Recommendations

We recommend that the Assistant Administrator for Research and Development:

2-1   Develop one or more measures linked to the short-term outcomes identified in the LRP MYP.

2-2   Augment the LRP's citation analysis measures (PART Measures 1 and 2) with measures that are meaningful to ORD program managers, and that are linked specifically to the LRP's goals and objectives.

2-3   Develop an implementation plan for the LRP client survey to:

> ➤ Identify the universe of LRP clients,
> ➤ Randomly select an appropriate target population,
> ➤ Conduct a representative survey of LRP clients, and
> ➤ Obtain a statistically valid response rate.

If ORD decides not to use the client survey tool, then ORD should develop a reliable alternative mechanism for collecting customer feedback along with an implementation plan for the alternative mechanism.

2-4   Provide BOSC with the following performance measurement data prior to full program reviews: (1) the results of the most recent client survey (or its alternative mechanism for collecting client feedback), (2) data sufficient to assess LRP's progress towards achieving program goals and outcomes, and (3) other data needed to support each of BOSC's peer review charge questions.

2-5   Require that BOSC program review reports include an explicit discussion of the reliability and suitability of the performance data that ORD provided to BOSC for each charge question and factor considered.

2-6   Revise ORD's guidance to BOSC for LTG ratings to ensure that all aspects of the summary assessment charge questions are clearly linked to the qualitative ratings definitions.

2-7   Supplement the current general LTG ratings definitions with program-specific milestones, and benchmarks for success, that are linked to elements in the LTG ratings definitions.

Agency Comments and OIG Evaluation

The Agency generally agreed with our recommendations and stated that ORD is actively incorporating these recommendations into the LRP program and, where appropriate, activities relevant to all BOSC research program reviews. However,

ORD responded that for recommendations closely linked to the OMB PART (recommendations 2-2, 2-6, and 2-7), it is awaiting additional guidance from OMB before proposing specific corrective actions. While we understand the need for ORD to follow OMB guidance, existing OMB guidance does not preclude ORD from augmenting its measures with more meaningful measures. Developing these measures in the interim would provide ORD with an opportunity to field test its measures before OMB revises its guidance and demonstrate the benefits of such measures as potential future measures. In our view, ORD should take action to implement recommendations 2-2, 2-6, and 2-7 in the interim. A summary of the Agency's response to each recommendation and our analysis follows.

Recommendation 2-1: ORD agreed to develop procedures, which the LRP NPD and laboratory line management will follow, to produce one or more measures linked to the short-term outcomes. ORD also responded that it will publish a report on recent outcomes of the LRP. Subsequent to ORD's response to the draft report, we requested and received clarification from ORD that it will develop short-term outcome measures for LRP by January 2011. We accept ORD's planned actions and the timeline for completion of the recommendation.

Recommendation 2-2: ORD responded that it will develop measures that are meaningful to ORD's program managers, pending OMB guidance. While we understand the need for ORD to follow OMB guidance, existing OMB guidance does not preclude ORD from augmenting its current citation analysis measures with more meaningful measures. We consider recommendation 2-2 to be open and unresolved.

Recommendation 2-3: ORD stated that it will develop a plan for obtaining LRP partner feedback. The plan will be completed by February 2011. We accept ORD's planned actions and the timeline for completion of the recommendation.

Recommendation 2-4: ORD responded that it will provide BOSC with the information listed (results of the most recent client survey or alternative; data on LRP's progress in achieving program goals and outcomes; and data to support each peer review charge question) to inform future BOSC reviews as appropriate for each new research program review. Subsequent to ORD's response to the draft report, we requested and received clarification from ORD that it will provide performance information to BOSC prior to the research program reviews. We accept ORD's planned actions and the timeline for completion of the recommendation.

Recommendation 2-5: ORD stated that it will revise the BOSC Program Review Report Guidance document by June 2011 to include this recommendation (an explicit discussion of the reliability and suitability of the performance data for each charge question and factor considered). We accept ORD's planned actions and the timeline for completion of the recommendation.

Recommendations 2-6 and 2-7: ORD stated that it is waiting for OMB direction before revising or eliminating the BOSC rating measures. ORD further stated that if it continues rating LTGs, these recommendations will be addressed. As long as ORD continues to have these LTG ratings, we continue to believe that ORD should ensure that all aspects of the summary assessment charge questions are clearly linked to the qualitative ratings definitions. Further, we believe that ORD should supplement the current general LTG ratings definitions with program-specific milestones, and benchmarks for success, that are linked to elements in the LTG ratings definitions. We consider Recommendations 2-6 and 2-7 to be open and unresolved.

The Agency also provided clarification comments related to the draft report. We made changes to the final report based on these comments, as appropriate. The Agency's complete written response is in Appendix D. Our evaluation of the Agency's response is in Appendix E.

Status of Recommendations and Potential Monetary Benefits

		RECOMMENDATIONS				POTENTIAL MONETARY BENEFITS (in $000s)	
Rec. No.	Page No.	Subject	Status[1]	Action Official	Planned Completion Date	Claimed Amount	Agreed To Amount
2-1	22	Develop one or more measures linked to the short-term outcomes identified in the LRP MYP.	C	Assistant Administrator for Research and Development			
2-2	22	Augment the LRP's citation analysis measures (PART Measures 1 and 2) with measures that are meaningful to ORD program managers, and that are linked specifically to the LRP's goals and objectives.	U	Assistant Administrator for Research and Development			
2-3	22	Develop an implementation plan for the LRP client survey to: ➤ Identify the universe of LRP clients, ➤ Randomly select an appropriate target population, ➤ Conduct a representative survey of LRP clients, and ➤ Obtain a statistically valid response rate. If ORD decides not to use the client survey tool, then ORD should develop a reliable alternative mechanism for collecting customer feedback along with an implementation plan for the alternative mechanism.	C	Assistant Administrator for Research and Development			
2-4	22	Provide BOSC with the following performance measurement data prior to full program reviews: (1) the results of the most recent client survey (or its alternative mechanism for collecting client feedback), (2) data sufficient to assess LRP's · progress towards achieving program goals and outcomes, and (3) other data needed to support each of BOSC's peer review charge questions.	C	Assistant Administrator for Research and Development			
2-5	22	Require that BOSC program review reports include an explicit discussion of the reliability and suitability of the performance data that ORD provided to BOSC for each charge question and factor considered.	C	Assistant Administrator for Research and Development			
2-6	22	Revise ORD's guidance to BOSC for LTG ratings to ensure that all aspects of the summary assessment charge questions are clearly linked to the qualitative ratings definitions.	U	Assistant Administrator for Research and Development			
2-7	22	Supplement the current general LTG ratings definitions with program-specific milestones, and benchmarks for success, that are linked to elements in the LTG ratings definitions.	U	Assistant Administrator for Research and Development			

[1] O = recommendation is open with agreed-to corrective actions pending
C = recommendation is closed with all agreed-to actions completed
U = recommendation is undecided with resolution efforts in progress

Details on Scope and Methodology

Our evaluation focused on the performance measures used by ORD's Land Research Program. However, the issues we identified, and our recommendations to address those issues, may be applied to other ORD research programs. We selected the Land Research Program, in part, because the program had not received a full BOSC review since 2006 and was not scheduled for one in the near future. We reviewed ORD guidance, policies, and procedures for performance measurement to identify ORD's guidelines pertaining to establishing effective research performance measures. We reviewed relevant literature (see the list at the end of this appendix) to identify established criteria for evaluating the effectiveness of performance measures. We evaluated LRP measures against these established criteria. We reviewed the most recent Bibliometric and Decision Document Analysis Report to determine any recent innovative developments which might be used to measure the effectiveness of ORD research programs. This report was completed for the Air Research Program in June 2009. We reviewed the supporting data provided by ORD to BOSC for the 2006 BOSC review.

We interviewed ORMA's staff, the NPD for Land Research, and the Designated Federal Officers for BOSC to identify performance measures used by the LRP, collect information on these measures, discuss current work on performance measurement, and discuss potential changes to existing measures. We contacted key LRP clients within EPA's Office of Solid Waste and Emergency Response, and within EPA regions that administer Resource Conservation and Recovery Act and Comprehensive Environmental Response, Compensation, and Liability Act programs, to obtain feedback on the effectiveness of the LRP client survey as a performance measurement tool. We also contacted BOSC Executive and Land Subcommittee members to obtain feedback related to their perspective of ORD policies and the adequacy of the information provided by ORD for BOSC reviews. We conducted our field work from May 2009 to May 2010.

Review of Management (Internal) Controls

Generally accepted government auditing standards require that auditors obtain an understanding of internal controls significant to the audit objectives and consider whether specific internal control procedures have been properly designed and placed in operation. We examined management and internal controls as they related to our objectives. Chapter 2 identifies findings and recommendations which may assist ORD in making better use of its resources that are devoted to performance measurement.

Sources for Establishing Performance Measurement Criteria

1. *Evaluating Research Efficiency in the U.S. Environmental Protection Agency.* Committee on Evaluating the Efficiency of Research and Development Programs at the U.S. Environmental Protection Agency, National Research Council of the National Academies. 2008.

2. Government Performance and Results Act of 1993 (GPRA).

3. *Measuring Performance at the Army Research Laboratory: The Performance Evaluation Construct.* Brown, Edward A., U.S. Army Research Laboratory's Special Projects Office, Army Research Laboratory Journal of Technology Transfer, Vol. 22 (2): 21-26 (June 1997).

4. *OMB's Program Assessment Rating Tool Guidance No. 2008-01* (Guidance). U.S. Office of Management and Budget. January 29, 2008.

5. *Performance Measurement Criteria Checklist.* Washington County, Oregon.

6. *Serving the American Public: Best Practices in Performance Measurement.* Benchmarking Study Report. The National Performance Review. June 1997.

7. The Performance-Based Management Handbook: A Six-Volume Compilation of Techniques and Tools for Implementing the Government Performance Results Act of 1993 (GPRA), Volume 2, *Establishing an Integrated Performance Measurement System.* Prepared by the Performance-Based Management Special Interest Group (PBM SIG) on a contract by the U.S. Department of Energy. September 2001.

   Included in the PBM SIG handbook:
   - Auditor General of Canada table for assessing the adequacy of performance measures
   - PBM SIG's Quality Check
   - PBM SIG's Three Criteria Check
   - University of California SMART test
   - U.S. Treasury Department criteria checklist

8. Various EPA Office of Inspector General reports including:

    EPA Performance Measures Do Not Effectively Track Compliance Outcomes, Report No. 2006-P-00006, December 15, 2005.

    Measuring the Impact of the Food Quality Protection Act: Challenges and Opportunities, Report No. 2006-P-00028, August 1, 2006.

    Performance Track Could Improve Program Design and Management to Ensure Value, Report No. 2007-P-00013, March 29, 2007.

    Using the Program Assessment Rating Tool as a Management Control Process, Report No. 2007-P-00033, September 12, 2007.

    Total Maximum Daily Load Program Needs Better Data and Measures to Demonstrate Environmental Results, Report No. 2007-P-00036, September 19, 2007.

Voluntary Programs Could Benefit from Internal Policy Controls and a Systematic Management Approach, Report No. 2007-P-00041, September 25, 2007.

Strategic Agricultural Initiative Needs Revisions to Demonstrate Results, Report No. 2007-P-00040, September 26, 2007.

Border 2012 Program Needs to Improve Program Management to Ensure Results, Report No. 08-P-0245, September 3, 2008.

Measuring and Reporting Performance Results for the Pollution Prevention Program Need Improvement, Report No. 09-P-0088, January 28, 2009.

EPA Needs a Comprehensive Research Plan and Policies to Fulfill its Emerging Climate Change Role, Report No. 09-P-0089, February 2, 2009.

9. U.S. Government Accountability Office report:

Measuring Performance: Strengths and Limitations of Research Indicators, United States General Accounting Office, GAO/RCED-97-91. March, 1997.

Criteria for Evaluating Performance Measures[26]

1) Well-Defined

    a. Is the measure clear, focused, and unambiguous to avoid misinterpretation?
    b. Is it clearly defined how the measure is relevant to the program?
    c. Is the information conveyed by the measure unique, or does it duplicate information provided by another measure?
    d. Can changes in the value of the measure be clearly interpreted as desirable or undesirable?
    e. Are data sources and specific requirements identified for the measure?
    f. Are any computations for the measure clearly specified?
    g. Are assumptions and definitions specified for what constitutes satisfactory performance?
    h. Does the measure include a clear statement of the end results expected?

2) Measurable, Quantifiable, and Comparable

    a. Is the measure objectively measurable, and based on observable information?
    b. Can the measure be quantified and compared to other data?
    c. Does the measure allow for comparison over time, or with other organizations, activities, or standards?
    d. Is the measure able to be compared to existing and past measures (i.e., is the measure able to show trends and define variation in performance?)?

3) Feasible

    a. Does the measure fit into the organization's resource constraints (i.e., budget, expertise, computer capability, etc.)?
    b. Is the measure cost-effective?[27]
    c. Are there privacy or confidentiality concerns that would prevent the use of the data by concerned parties?
    d. Is the measure doable within the time frame given?
    e. Are the data timely enough for evaluating program performance?

---

[26] The criteria identified in this appendix were compiled by OIG and are based on information obtained from the literature sources cited at the end of the appendix. The criteria were used to evaluate the appropriateness, or effectiveness, of performance measures used by ORD's Land Research Program.

[27] The measure should be available or able to be obtained with reasonable cost and effort, and provide maximum information per unit of effort. The cost of collecting data should not outweigh their value.

4) Consideration of Stakeholder Requirements and Feedback

    a. Does the measure support customer requirements?
    b. Has the measure been mutually agreed upon by the organization and its customers?
    c. Are the interests and expectations of the customer reflected in the measure?

5) Meaningful to Internal Stakeholders (managers, staff, etc.)

    a. Does the measure identify gaps between current status and performance aspirations, thereby highlighting opportunities for improvement?
    b. Do the measures enable strategic planning, and drive actions required to achieve objectives and strategic goals?
    c. Can management actions influence the results of the measure (i.e., does the measure show where management can take action to change the program results, and drive improvement for the given metric?)?
    d. Does the measure focus on the effectiveness and/or efficiency of the system being measured?
    e. Is the measure perceived as valuable by the organization and the people involved with the metric?
    f. Does the measure clearly reflect changes in the program?
    g. Does the measure cover an appropriate portion of the program's operations?
    h. Does the measure reflect the desired outcomes of the program?

6) Logical Design

    a. Does the measure include milestones and/or indicators to express qualitative criteria?
    b. Are there reliable benchmark data, standards, or alternative frames of reference for interpreting the measure?
    c. Is the measure clearly attributable to specific program activities?
    d. Have appropriate industry or other external standards been applied to the measure?
    e. For survey data, have the survey questions and survey methodology been prepared, or at least reviewed by, professionals with demonstrated survey research qualifications?

7) Functional

    a. Does the measure encourage the right kind of behavior (i.e., does the measure align behavior with the program's strategy and organizational priorities?)?
    b. Is the measure vulnerable to producing unintended consequences?

8) Reliable and Available Data

    a. Are data available, or can they be collected, for the measure?
    b. Can the data required for the measure be replicated (i.e., will repeated measurements yield consistent results?)?
    c. Are the data for the measure susceptible to biases, exaggerations, omissions, or errors that are likely to make the measure inaccurate or misleading?
    d. Are data samples for the measure large enough to yield reliable data within acceptable confidence limits?
    e. Is there a clear audit trail for the measure that would allow for tracing the measure back to the detailed data used to compile the measure?
    f. If the measure is not quantitative, is it reasonable to verify it through an audit or review by an expert panel?

9) Connection with Program Goals and Objectives

    a. Is the measure clearly linked to the programs goals and objectives?
    b. Does the measure provide a clear understanding of progress toward objectives and strategic goals (i.e., an indication of current status, rate of improvement, and probability of achieving the objectives and goals)?

Sources for Criteria Identified Above

1) *Developing and Presenting Performance Measures for Research Programs.* Research Roundtable, August 1, 1995.

2) EPA Office of Inspector General (various reports).

3) *Evaluating Research Efficiency in the U.S. Environmental Protection Agency.* Committee on Evaluating the Efficiency of Research and Development Programs at the U.S. Environmental Protection Agency, National Research Council of the National Academies, 2008.

4) Government Performance and Results Act of 1993 (GPRA).

5) *Measuring Performance at the Army Research Laboratory: The Performance Evaluation Construct.* Brown, Edward A., U.S. Army Research Laboratory's Special Projects Office, Army Research Laboratory Journal of Technology Transfer, Vol. 22 (2): 21-26 (June 1997).

6) *OMB's Program Assessment Rating Tool Guidance No. 2008-1* (Guidance). U.S. Office of Management and Budget, January 29, 2008.

7) *Performance Measurement Criteria Checklist.* Washington County, Oregon.

8) *Serving the American Public: Best Practices in Performance Measurement.*
Benchmarking Study Report. The National Performance Review, June 1997.

9) The Performance-Based Management Handbook: A Six-Volume Compilation of
Techniques and Tools for Implementing the Government Performance Results Act of
1993 (GPRA), Volume 2, *Establishing an Integrated Performance Measurement System.*
Prepared by the Performance-Based Management Special Interest Group (PBM SIG) on a
contract by the U.S. Department of Energy, September 2001.

Included in the PBM SIG handbook:

- Auditor General of Canada table for assessing the adequacy of performance measures
- PBM SIG's Quality Check
- PBM SIG's Three Criteria Check
- University of California SMART test
- U.S. Treasury Department criteria checklist

10) U.S. Government Accountability Office (various reports).

Linking Summary Assessment Charge Questions to LTG Qualitative Ratings Definitions

<u>Exceptional:</u> indicates that the program is meeting all and exceeding some of its goals, both in the quality of the science being produced and the speed at which research result tools and methods are being produced. An exceptional rating also indicates that the program is addressing the right questions to achieve its goals. The review should be specific as to which aspects of the program's performance have been exceptional.

Summary Assessment Charge Question 1	"An exceptional rating also indicates that the program is addressing the right questions to achieve its goals."
Summary Assessment Charge Question 2	"…meeting all and exceeding some of its goals, both in the quality of the science being produced…"
Summary Assessment Charge Question 3	Not addressed by definition

<u>Exceeds Expectations:</u> indicates that the program is meeting all of its goals. It addresses the appropriate scientific questions to meet its goals and the science is competent or better. It exceeds expectations for either the high quality of the science or for the speed at which work products are being produced and milestones met.

Summary Assessment Charge Question 1	"It addresses the appropriate scientific questions to meet its goals…"
Summary Assessment Charge Question 2	"…the science is competent or better. It exceeds expectations for either the high quality of the science …"
Summary Assessment Charge Question 3	Not addressed by definition

<u>Meets Expectations:</u> indicates that the program is meeting most of its goals. Programs meet expectations in terms of addressing the appropriate scientific questions to meet their goals, and work products are being produced and milestones are being reached in a timely manner. The quality of the science being done is competent or better.

Summary Assessment Charge Question 1	"Programs meet expectations in terms of addressing the appropriate scientific questions to meet their goals …"
Summary Assessment Charge Question 2	"The quality of the science being done is competent or better."
Summary Assessment Charge Question 3	Not addressed by definition

<u>Not Satisfactory:</u> indicates that the program is failing to meet a substantial fraction of its goals, or if meeting them, that the achievement of milestones is significantly delayed, or that the questions being addressed are inappropriate or insufficient to meet the intended purpose. Questionable science is also a reason for rating a program as unsatisfactory for a particular long-term goal. The review should be specific as to which aspects of a program's performance have been inadequate.

Summary Assessment Charge Question 1	"…the questions being addressed are inappropriate or insufficient to meet the intended purpose."
Summary Assessment Charge Question 2	"Questionable science is also a reason for rating a program as unsatisfactory for a particular long-term goal."
Summary Assessment Charge Question 3	Not addressed by definition

Source: OIG analysis of ORD guidance

Agency Response to Draft Report

UNITED STATES ENVIRONMENTAL PROTECTION AGENCY
WASHINGTON, D.C. 20460

OFFICE OF
RESEARCH AND DEVELOPMENT

JUNE 1 7 2010

<u>**MEMORANDUM**</u>

SUBJECT: Office of Research and Development's (ORD) Response to OIG Draft Report
EPA's Office of Research and Development Performance Measures Need Improvement,
Assignment No. FY2009-0891

FROM: Paul T. Anastas /s/
Assistant Administrator

TO: Wade T. Najjum
Assistant Inspector General for Program Evaluations

      Thank you for the opportunity to comment on the Office of Inspector General (OIG) draft audit report, *EPA's Office of Research and Development Performance Measures Need Improvement* (Assignment No. FY2009-0891), dated May 11, 2010, which focused on ORD's Land Research Program (LRP). The recommendations provided in the draft audit report will help ORD continue to improve its performance measures.

      As the scientific research and assessment arm of EPA, ORD recognizes the complexity of assessing research performance and your efforts to integrate an understanding of that complexity into this report. Our commitment to ensuring that our science is of the highest quality and our programs are managed effectively and efficiently means we are continually searching for new and improved methods by which to measure the performance of our research programs.

      ORD has identified areas in the draft report that we would like to clarify:

1.   We recognize that the LRP's measures can be improved. However, until the Office of Management and Budget (OMB) provides new performance guidance, ORD continues to report on its measures developed during the Program Assessment Rating Tool (PART) process.

See Appendix E, Note 1 for OIG Response.

2. We agree that the Survey of LRP clients did not provide a meaningful measure of customer feedback. However, it should be noted that the LRP survey was the first survey of its kind, and the results of the process were used to develop ORD guidance for subsequent surveys.

See Appendix E, Note 2 for OIG Response.

3. ORD's matrix management structure is noted several times for having a negative impact upon developing performance measures. The matrix management structure has, in fact, assisted ORD in managing the complex nature of research being conducted in numerous laboratories and centers across the country.

See Appendix E, Note 3 for OIG Response.

The OIG provides seven recommendations to strengthen ORD's LRP program. In general, we agree with the recommendations, and I am pleased that ORD is actively incorporating these recommendations into the LRP program and, where appropriate, activities relevant to all Board of Scientific Counselors research program reviews. However, for recommendations that are closely linked to the OMB PART, we are awaiting additional guidance from OMB before proposing specific corrective actions.

See Appendix E, Notes 4 -10 for OIG Responses to ORD's Proposed Actions.

Attached please find a summary table of ORD's corrective actions and associated projected completion dates. If you have any questions, please contact Deborah Heckman at (202) 564-7274.

Attachment

cc: Lek Kadeli
Kevin Teichman
Amy Battaglia
Desmond Mayes
Deborah Heckman
Randall Wentsel
Greg Susanke
Alvin Edwards

ORD Corrective Actions and Projected Completion Dates

Rec No.	OIG Recommendation	Lead Responsibility	ORD Corrective Action	Planned Completion Date
2-1	Develop one or more measures linked to the short-term outcomes identified in the LRP MYP.	Assistant Administrator for Research and Development	ORD will develop procedures, which the LRP National Program Director (NPD) and laboratory line management will follow, to produce one or more measures linked to short-term outcomes.	January 2011
			In addition, ORD will publish a report on recent outcomes of the LRP.	September 2010
2-2	Augment the LRP's citation analysis measures (PART Measures 1 and 2) with measures that are meaningful to ORD program managers, and that are linked specifically to the LRP's goals and objectives.	Assistant Administrator for Research and Development	ORD will develop measures that are meaningful to ORD's program managers, pending OMB guidance.	Pending OMB performance guidance
2-3	Develop an implementation plan for the LRP client survey to: ➤ Identify the universe of LRP clients, ➤ Randomly select an appropriate target population, ➤ Conduct a representative survey of LRP clients, and ➤ Obtain a statistically valid response rate; or If ORD decides not to use the client survey tool, then ORD should develop a reliable alternative mechanism for collecting customer feedback along with an implantation plan for the alternative mechanism.	Assistant Administrator for Research and Development	ORD will develop a plan for obtaining LRP partner feedback.	February 2011

2-4	Provide Board of Scientific Counselors (BOSC) with the following performance measurement data prior to full program reviews: (1) the results of the most recent client survey (or its alternative mechanism for collecting client feedback), (2) data sufficient to assess LRP's progress towards achieving program goals and outcomes, and (3) other data needed to support each of BOSC's peer review charge questions.	Assistant Administrator for Research and Development	ORD will provide the BOSC with the requested information listed to inform future BOSC Research Program Reviews.	As appropriate for each new Research Program Review
2-5	Require that BOSC program review reports include an explicit discussion of the reliability and suitability of the performance data that ORD provided to BOSC for each charge question and factor considered.	Assistant Administrator for Research and Development	ORD will revise the BOSC Program Review Report Guidance document to include this recommendation.	June 2011
2-6	Revise ORD's guidance to BOSC for LTG ratings to ensure that all aspects of the summary assessment charge questions are clearly linked to the qualitative ratings definitions.	Assistant Administrator for Research and Development	ORD is waiting for OMB direction before revising or eliminating the BOSC rating measures. If ORD continues rating Long-Term Goals (LTG), this recommendation will be addressed. ORD finds BOSC responses to the charge questions and their recommendations to be very beneficial in improving research program performance. However, the BOSC's ratings are not as useful in improving program performance. Progress ratings for each LTG (i.e., exceptional, exceeds expectations, meets expectations, not satisfied) were developed to satisfy OMB PART requirements. Furthermore, the BOSC spends a considerable amount of time deliberating over the LTG ratings. ORD prefers that the BOSC spend its time on addressing the charge questions and providing ORD with recommendations for improving its research programs. Pending OMB drivers will inform ORD guidance with respect to the ratings.	Pending OMB performance guidance

2-7	Supplement the current general LTG ratings definitions with program-specific milestones, and benchmarks for success, that are linked to elements in the LTG ratings definitions.	Assistant Administrator for Research and Development	ORD will wait for OMB direction before revising or eliminating the BOSC rating measures. If ORD continues rating LTGs, this recommendation will be addressed. See ORD's response to 2-6 for explanation.	Pending OMB performance guidance

OIG Evaluation of Agency Response

Note 1 - We stated on pages 8 and 14 that ORD is waiting on guidance from OMB before changing its PART measures. However, we believe that ORD should augment the current measures with additional measures before guidance from OMB is issued. After new OMB guidance is issued, then ORD should propose these new measures to replace the current PART measures, if appropriate, considering the new guidance.

Note 2 - We added the following sentence in Chapter 2 under the heading *ORD's Survey of LRP Clients Did Not Provide a Meaningful Measure of Customer Feedback*: "According to ORD, the LRP survey was the first survey of its kind, and results of the process were used to develop ORD guidance for subsequent surveys."

Note 3 - In its response to our draft report, ORD stated that ORD's matrix management structure is noted several times for having a negative impact on developing performance measures. According to ORD, the matrix management structure has assisted ORD in managing the complex nature of research being conducted in numerous laboratories and centers across the country. However, while the matrix management structure may assist in managing some aspects of research, our report focused only on the effectiveness of performance measures for the LRP at the time of our review. We found that the effectiveness of the LRP's performance measures was impacted by the management structure employed at ORD.

Note 4 - We appreciate ORD's commitment to publish a report on outcomes of the LRP by September 2010, and to develop procedures for producing measures linked to short-term outcomes. However, it was not clear from the response whether ORD intended to issue guidance by January 2011, or develop new performance measures by that date. Subsequent to ORD's response to the draft report, we requested and received written clarification on July 19, 2010, that ORD will develop short-term outcome measures for LRP by January 2011. With this clarification, we accept ORD's planned actions and the timeline for completion of Recommendation 2-1.

Note 5 - Regarding Recommendation 2-2, we continue to believe that ORD should augment its current PART measures with additional measures that are meaningful to ORD program managers, and that are linked specifically to the LRP's goals and objectives. ORD can still report on the existing PART measures, as well as the new measures it develops. This interim approach would allow ORD to field test the new measures and demonstrate the benefits of such measures as potential future measures before guidance from OMB is issued.

Note 6 - We accept ORD's planned actions and the timeline for completing Recommendation 2-3.

Note 7 - We accept ORD's planned actions to provide BOSC with the information listed (results of the most recent client survey or alternative; data on LRP's progress in achieving program goals and outcomes; data to support each peer review charge question) as appropriate for each new research program review. However, the intent of our recommendation was that the information would be provided to BOSC prior to its program reviews. Subsequent to ORD's response to the draft report, we requested and received written clarification on July 19, 2010, that ORD will provide the information to BOSC prior to its program reviews. With this clarification, we accept ORD's planned actions and the timeline for completion of Recommendation 2-4.

Note 8 - We accept ORD's planned actions and the timeline for completing Recommendation 2-5.

Note 9 - Regarding Recommendation 2-6, our recommendation is to improve the guidance to BOSC, not to revise or eliminate the LTG ratings. As long as ORD continues to have these LTG ratings, we continue to believe that ORD should ensure that all aspects of the summary assessment charge questions are clearly linked to the qualitative ratings definitions.

Note 10 - Regarding Recommendation 2-7, our recommendation is to supplement the guidance to BOSC with performance information that better enables BOSC to assign LTG ratings, not to revise or eliminate the LTG ratings. As long as ORD continues to have these LTG ratings, we continue to believe that ORD should supplement the current general LTG ratings definitions with program-specific milestones, and benchmarks for success, that are linked to elements in the LTG ratings definitions.

Distribution

Office of the Administrator
Assistant Administrator for Research and Development
Agency Follow-up Official (the CFO)
Agency Follow-up Coordinator
Deputy Assistant Administrator for Management, Office of Research and Development
Deputy Assistant Administrator for Science, Office of Research and Development
General Counsel
Associate Administrator for Congressional and Intergovernmental Relations
Associate Administrator for External Affairs and Environmental Education
Audit Follow-up Coordinator, Office of Research and Development
Inspector General